HAMLYN - COLOURFAX

EXPLORING

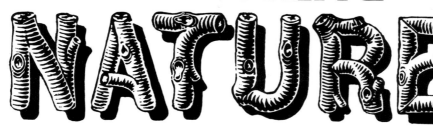

Jim Flegg

CONTENTS

HAMLYN

cagoule/anorak with hood   hat

Wildlife is all about us, wherever we are. The greatest delight of exploring nature, even if you have only a few spare moments, is that the wonders of nature are always there for you to explore. The basic skills needed are simple – with enthusiasm, a little patience, time and practice, anyone can become an expert. *Exploring Nature* takes you through the realms of animals and plants, and through various different habitats, showing you just a fraction of what is there to interest you. It explains this complex world, giving many examples of the perfect adaptation of plant or animal to its way of life or its habitat.

Above: *Remember that a hand lens or magnifying glass will prove invaluable. Without one you may miss a wealth of detail, particularly in the worlds of insects and plants.*

The equipment you will need depends very much on what interests you. Bird watchers must have binoculars. Any naturalist, and in particular plant or insect watchers, must have a hand lens. These allow you to see easily the best of nature, revealing a world the unaided eye cannot fully enjoy. The equipment you will need for each type of nature watching is explained in more detail in each chapter.

binoculars   wellies   warm clothes

Above: *Nature watching demands layers of sensible weatherproof clothes, not too loud or brightly coloured. Remember, you can always take something off if you are too hot!*

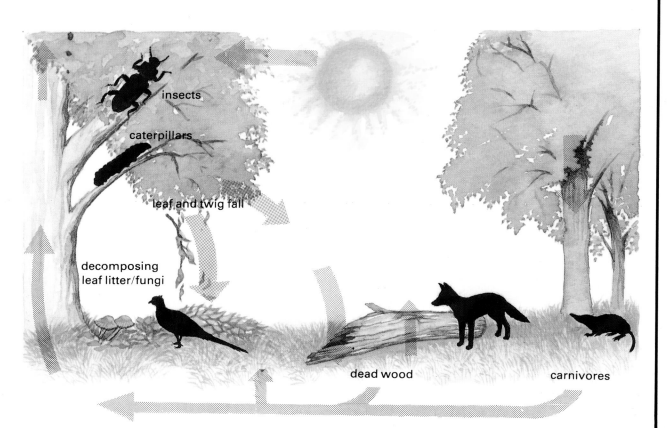

The diagram labels, reading around the cycle:
insects
caterpillars
leaf and twig fall
decomposing leaf litter/fungi
dead wood
carnivores

THE ENDLESS FOOD CHAIN

Within any habitat, the animals and plants form part of a web of life. Energy from sunlight provides the power that makes plants grow. Some animals (herbivores) feed directly on plants.

Carnivores (meat eaters) come in all sizes, as do their prey. So there are mammal predators as big as a whale, and insect and mite predators smaller than a pinhead.

Falling leaves, dead branches, and animal dung are decomposed by fungi and microbes. This enriches the soil which in turn helps plants grow, completing the cycle.

Date - 25 July
Place - Paulton Pool
Weather - Cool damp

Left: *No matter which aspects of nature you are watching, noting and sketching what you see is essential. Without these to remind you, many details would be soon forgotten.*

woodland

fields

Left: Woodland glades full of wild flowers make an ideal sheltered territory for butterflies. This is a speckled wood butterfly.

Right: Although the fox is a typical farmland predator, it is adaptable enough to enjoy life in many urban areas too.

Wildlife is all around us, but nature has her specialists just as we do. The animals and plants of these four habitats shown above are not the same, although a few particularly adaptable species may be found almost anywhere.

Woodland, with its large, long-lived trees, tends not to change much from year to year. It is sometimes described as a place with a stable environment, where you can expect to find the same sorts of plants and animals every year.

There are differences between woodland types – particularly between the deciduous woods which lose their leaves in winter and coniferous woods, where the trees are normally evergreen.

Fields change dramatically during the year as different crops grow and are harvested and the soil is ploughed ready for the next crop. These are unstable environments, but the hedges and copses around them are often rich in plants and animals.

Uplands are harsh but stable environments, often wet and windy in summer, and usually very cold in winter. Most plants and animals in the uplands are specialized to cope with these conditions, and the wildlife is not as varied as in other places.

Coastal habitats, too, are harsh – not so much because they are cold, but because of the salt and the action of the waves. Here too, plants and animals tend to be specialized.

uplands

coastal habitats

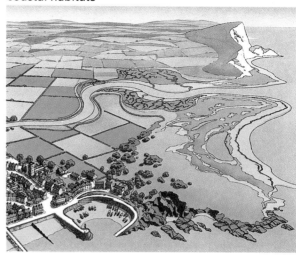

Left: Ravens are amongst the few birds that can live all year in the harsh weather of upland areas.

Right: Look for crabs on sandy shores as well as under seaweed and in rock pools. This is a common shore crab.

NATURE RESERVES

Throughout Britain and Europe there is a network of nature reserves belonging to various conservation societies. They allow nature watchers to see at their best a selection of habitats and their special plants and animals. Permanent hides offer an excellent view without causing any disturbance. They are ideal for photographers.

Binoculars are essential for a birdwatcher. With them, distant black dots take on identifiable colours and shapes – without them, they remain just black dots! Prices vary greatly, but a reasonable pair can be bought quite cheaply. Go to a good local shop to try them out before buying, because they must be easy and comfortable for *you* to use.

Real enthusiasts will find a telescope most useful in places such as estuaries, large lakes and reservoirs, particularly when they are looking at ducks, geese and waders. This is because the telescope's magnification is that much greater, and in these areas you usually remain at some distance from the birds you are watching. A pocket-sized "field guide" identification book is very helpful, but remember always to make your own notes and sketches of the birds as you saw them, with details of date, place and weather conditions.

Left: *Binoculars with ×8 or ×9 magnification are suitable for most habitats. It is sometimes better to use ×10 by the sea, but remember that these will be heavier.*

Left: *A good telescope gives superb views of distant birds, and a tripod is essential to keep it steady. Unfortunately both are expensive.*

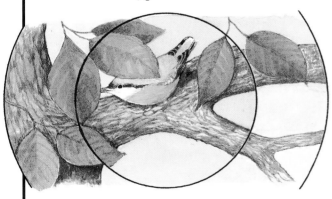

Above: *Using a telescope well takes practice, but the rewards are great.*

DO'S AND DON'TS OF BIRDWATCHING

√ keep alert
√ take time to stop look and listen

× hurry
× make a loud or sudden noise
× wear bright clothing

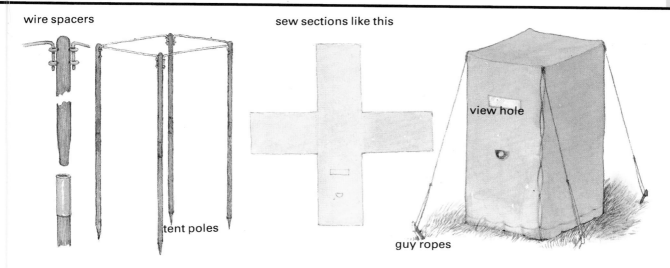

wire spacers

sew sections like this

view hole

tent poles

guy ropes

APPROACHING BIRDS

Of all animals, birds are often the most alert and wary. The rule when birdwatching is therefore to do your best to see the bird before it sees you. A stealthy approach is needed, moving quietly, with frequent pauses to **listen** – birds can often be heard singing or calling long before you catch sight of them.

Take a route that keeps you in the shelter of nearby bushes if you are out in the open, and never walk along the top of a bank or sea wall. Keep below the skyline, putting your head up cautiously every so often to check what is there. Another technique is to use a hide – but remember that you must not disturb the birds, for their sake, and that nesting birds are protected by law.

MAKING A HIDE

A portable hide is really a small square tent, usually of canvas or camouflage material, with an observation slit in one or more sides. Tent poles and stiff wire spacers make a sturdy frame, over which the canvas cover fits. The cover should be sewn together as a cross. The entrance flap and observation slit are closed with zips or "velcro" fastenings. The tent is kept in position by guy ropes and tent pegs hold the canvas down to prevent flapping.

The hide should be small enough to be easily carried, but large enough to take you, a seat (an absolute must, or you will move to ease a stiff leg and disturb the birds), your binoculars and perhaps a tripod and camera or telescope.

Above: *Hides are not difficult to make – or you could buy one. Don't forget that cars make good "hides" for bird watching.*

Right: *Ringing birds, to monitor their movements, is rewarding work. Join a local birdwatching society, and you may be able to take part.*

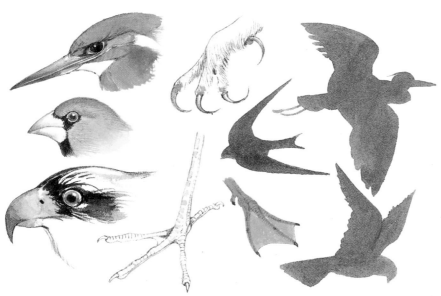

Left: *Look for and note down any distinguishing features of the birds you see, as you see them. The size, colour and shape of wings, beak and feet will all help to identify them later. They will also give you a lot of clues about the birds, particularly about their feeding habits.*

Below: *Note carefully how the birds you have spotted fly. Straight, swooping, soaring or hovering flight patterns will help in identification.*

Birds have a number of features that help them survive in their chosen habitat, and these characteristics will help you sort out just what group the bird belongs to. Good examples are the powerful hooked beaks of the birds of prey, the dagger-like beaks of herons and the kingfisher, the webbed feet of ducks and geese and the longer legs of herons and waders. Seed-eaters (such as finches) have short, stout, triangular beaks, while fruit and insect-eaters (thrushes and warblers) have medium-length slim beaks. Birds of the open skies, such as the swift, have long slender wings, while woodland birds, such as the dove, have short wings.

Below: *Note down as much as you can about any birds you see while you are out. A few rough drawings will help, too. Details are easily forgotten once you get home!*

25th June

dark wing

long beak

light breast

- Britain's longest-haul migrant is the Arctic tern. It breeds in the Arctic but flies south to winter off the Antarctic ice-cap, a round trip of about 30,000 kilometres.

- At around 35 kilograms, the mute swan (Britain's heaviest bird) is about 7,000 times heavier than the 5 gram goldcrest, the lightest.

- Britain's commonest nesting birds are the house sparrow, starling and chaffinch. There are more than 5 million pairs of each at peak times.

- Britain's rarest bird is the snowy owl. Since 1983 7 females have been sighted, but no males.

Above: *Most estuary birds have to probe deep into the mudflats at low tide for worms and shellfish, so many have long beaks. They usually have long legs (so that they don't get wet!) and all have long toes to stop them sinking into the mud.*

USING FIELD GUIDES

In field guides, birds are often grouped in families of related species (ducks, hawks, waders, and so on). Once you have checked a bird's distinguishing features, such as beak, legs and feet, you will be able to work out which family you think it belongs to. Then you can turn to the group of illustrations to see which bird best fits your notes.

Some guides are designed for use only in one habitat – woodland or marshland for example. This can be helpful, as it greatly reduces the possibilities of confusion, but never forget that birds fly between habitats or have a wide habitat choice. "Sea" gulls, for example, may commonly be found on a grassy field.

As you go birdwatching through the different seasons of the year and you visit new habitats, you will build up a store of useful identification features that help you take short-cuts. The way a bird moves or flies (called its *jizz*) may tell you more than the colours of its feathers, and more quickly. Songs and calls, too, are not only enjoyable to listen to, but useful aids to identification. Again, experience helps greatly, but you can learn a great deal by listening to the tapes or records of bird song now readily available. Try your local library.

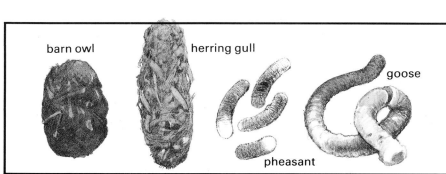

barn owl   herring gull   pheasant   goose

Left: *Pellets or droppings provide good clues as to what birds eat – mouse bones and fur are often found in owl pellets, fishbones in a gull's. Pheasant droppings often contain grain husks, goose droppings grass fragments.*

Right: *Badgers spend most of the daylight hours in an underground tunnel network called a sett.*

Right: *For winter warmth and shelter, squirrels make a drey of twigs and leaves high in a tree.*

Below left: *The plump water-vole's burrow, deep in the river bank, sometimes has a secret underwater entrance.*

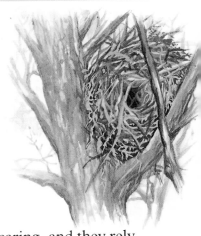

and an acute sense of hearing, and they rely on these two senses rather more than on their eye-sight. Moving quietly and carefully is extremely important!

Stalking wary mammals such as deer is never easy because they are well protected by their acute senses. It is best to walk *upwind* (you should move into the wind) to reduce noise and remove your scent, carefully checking the wind direction first by raising a licked finger. Before you set out check that your clothing is as dull-coloured as possible and

- Bats are the only mammals that really fly. They use in-flight echo-location to avoid obstacles and to identify their insect prey after dark.

- Dormice are our only truly hibernating mammals: they may sleep from November to May.

- The fastest breeding mammal in Britain is the meadow vole. It can reproduce from the age of 25 days and can have up to 17 litters of 6–8 young in a year.

"Mammal watching" is exciting and rewarding, but more difficult than birdwatching. Often there are only a few species to be found in one habitat and many mammals are more active at night than during the day. Although many have good daytime vision, all have an exceptionally keen sense of smell. Most have large, moveable ears (to gather in sounds)

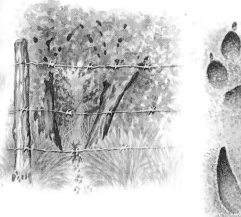

fox

deer

Above: *Deer often trample the undergrowth and damage trees with their antlers in the mating season.*

Above: *A lodge of logs like a dam is a sure sign of beavers.*

Above right: *Many mammals – particularly hedgehogs, badgers, foxes and deer – use the same trails regularly.*

dog   hare   fox

Left and above: *Most mammals leave distinctive tracks, good clues to their presence. Look out for them and you will discover a fascinating new interest. The long "slots" of deer tracks are distinctive, as are the "two long, two short" footprints of rabbits and hares. Fox and dog are more difficult to separate.*

when animals are spotted, use every scrap of natural cover available. Once you are really close you will need to lie flat and wriggle like a snake. Hard work – but the close-up views will make it worth while!

TRACKS AND SIGNS

One of the most exciting ways of finding out about the presence of mammals is by looking for their signs. There are all sorts of things to look out for, including food remains, tracks in snow or mud, droppings and pathways through the woods. All these signs can be checked to tell you which mammal made them.

Tracks are easily spotted throughout the year, especially in wet mud or sand. Check for clear-cut animal paths through grass, or tufts of hair on barbed wire.

Discarded nut-shells are a good find, since each animal opens nuts in a different way. Squirrels split them lengthways into two halves, while mice and voles gnaw holes in the shell. Keep your eyes open – for the possibilities for detection are endless!

LOOKING AT FLOWERS

There is little in the world of nature to rival the beauty of an old meadow, rich in flowering plants in full bloom in midsummer. Sadly, today's farming technology makes such a sight more and more rare over much of Britain and Europe, except on the poor farming lands of the west and in Alpine meadows.

If you look at a meadow full of flowers you will be struck by a kaleidoscope of patterns, shapes and colours. But a hand lens will allow you to see many finer points of detail, so that you can identify each flower correctly. Flower field guides are sometimes arranged by colour, but more often by family. A little experience will allow you to place an unknown plant in, for example, the pea family, or the roses, or the "umbellifers" (wild carrot, hogweed and so on, with umbrella-shaped flower heads) because of its flower shape. Once you have been able to work out which family a flower belongs to, it is then not hard to identify the species.

Above: *Flowers spend their lives producing seeds. These seeds are very important, for they are the beginnings of new plants. They are spread out or scattered in four main ways: they are shot out by the parent plants, or are carried away by wind, water or animals. A wide variety of insects are used of which the bees and flies are the best known. Besides their bright, insect-attracting colours, some flowers also have a supply of sugar-rich nectar, that bees turn into honey.*

1

2

3

HOW FLOWERS REPRODUCE

Flowering plants reproduce by means of seeds. Some may be blown long distances, as in the case of the downy "parachutes" of thistles. The musk or monkey-flower seed pod explodes violently when touched, scattering seeds far and wide. The well-named cleavers have hooked seeds that cling to animal fur (and human clothing!) while the wild rose and yew have tempting berries that birds eat, the seeds passing out in their droppings later – and far away.

Left: *Heather (1) indicates peaty soil, bee orchids (2) chalky soil and herb Paris (3) ancient woodland.*

Right: *Hedgerows are often rich in flowers – they are the last refuge of the plants that used to grow in the meadows.*

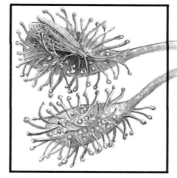

WHERE PLANTS LIVE

Plants can tell us a great deal about the habitat in which they are found. The presence of some may suggest the possibility of other plants that enjoy similar conditions. Heather (or ling) for example, is typical of damp, peaty moorland soils that are acid in nature. In contrast, many orchids, such as the bee orchid favour chalky soils. (The flower of this particular orchid looks just like a bee and attracts male pollinator bees, which believe they have found a mate!) Herb Paris also grows in chalky soil, and its presence tells you that you are in very old woodland.

Above: *Some plants are carnivorous (meat-eaters). This sundew traps passing insects on its sticky leaves and digests them to obtain vital food.*

LOOKING AT INSECTS

red admiral
butterfly

Left: *Butterflies are among the most colourful insects.*

Below: *With his antennae, the male lappet moth scents the female hundreds of metres away.*

lappet moth

dragonfly

There are more species of insects in the world than any other animal – many *millions* of different species, and an unimaginably large number of individuals! Almost all insects have six legs and two antennae, and most have one or two pairs of wings and can fly very skilfully: hoverflies and some moths can hover and even fly backwards. Some seem fragile – damselflies for example – while others, such as the beetles, are extremely tough.

Many insects and their larval stages feed on plants, but there are also lots of predatory species – the well-known ladybird among them. Many insects are small and a hand lens is useful to see them better. The best way to study insects and their behaviour is to keep them in a large jar with a piece of gauze secured over the top.

Right: *Dragonflies often mate on the wing.*

hoverfly

The hoverfly looks like a wasp, so birds avoid it.

Right: *Water boatmen are aquatic bugs. Two legs are adapted as "oars".*

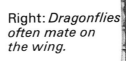

water boatman

Right: *Hoppers can leap enormous distances.*

hopper

CHECKLIST Equipment	Where to find insects
● sweep net	* in the air
● hand lens	* on plants, usually beneath leaves
● keeping tubes	* under decaying logs
● large jar	* on waterweed

Unlike mammals and birds, which have a bony skeleton inside their bodies, insects have an exoskeleton – a hard case. As they grow larger, they have to moult or shed it as a new one grows. This outer skeleton provides the insect with a firm base for fixing its muscles. It also stops the creature's body from drying out.

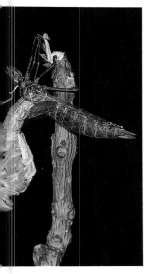

Above: *It sometimes takes several years for the the dragonfly nymph to grow to maturity underwater. Finally it climbs a reed stem into the open air and emerges*

from its cast skin, flying off to hunt as the elegant adult. The adult insect (to us, the dragonfly) may only represent a tiny proportion of the insect's life-cycle.

BLUFFS AND MIMICS

Keep an eye open for examples of insects that mimic unpleasant species. Some, the hoverfly for instance, mimic other insects, while others merge with their background to avoid being seen (and eaten!). A green caterpillar is difficult to see on a green leaf – a simple but effective form of self protection.

hoverfly

A LIFE CYCLE

Many insects start life as an egg. This hatches, and a larva emerges that moults several times, growing bigger at each stage. This changing shape is called metamorphosis. A caterpillar, for example, is the larva of a moth or butterfly. After a time the developing caterpillar forms a hard case round its body and stops moving. It is then known as a pupa. Later, the case splits open and out crawls an adult butterfly or moth.

USING A SWEEP NET

Sweep nets may be circular, semi-circular or triangular, roughly 30 cm across. They are best made from a stiff wire loop, bent to shape, then attached firmly to a broom handle 40 or 50 cm long. The net is made of "lace" curtain with a fine mesh, preferably soft so that it does not damage the insects, yet tough enough not to tear easily.

Insects can be chased in flight, or gathered by gently sweeping the net back and forth through plants, including grass. Try holding the net beneath a leafy tree branch and tapping the branch to dislodge insects into the net.

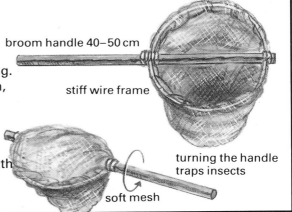

broom handle 40–50 cm

stiff wire frame

turning the handle traps insects

soft mesh

AMPHIBIANS AND REPTILES

There are relatively few amphibians and reptiles in Britain and Europe, compared with tropical parts of the world. But those few are well worth studying and some can be easily kept.

Amphibians, as their name implies, spend part of their lives in water, and part of it out on land. Commonest over much of Europe are the frogs and toads, familiar to everyone with their extremely powerful long back legs for hopping. Lizard-like in shape, but not related, are newts and salamanders, whose webbed feet and flattened tails indicate their abilities as swimmers. In Britain, some rare newts are protected by law and must not be touched.

Reptiles have a harder, drier skin than amphibians and (though many can and do swim) have no need to be close to water. They include the lizards, which are more numerous and more obvious in southern Europe and the snakes. The most common of these are the grass snake and adder, but remember the adder has a poisonous bite. **Watch** it, but **never** touch it. Snake-like, but actually a legless lizard, is the slow worm. All are cold-blooded, so are active only in the warmer summer months and during the heat of the day.

Most amphibians are commonly found in ponds or in damp shady places nearby. They can be kept in an aquarium which **must** have rocks emerging from the water. Lizards and snakes can be kept in a vivarium – a dry glass tank with sand, leaves and rocks at the bottom. Feed them on worms, flies, maggots and meal worms.

REPRODUCTION

Amphibians mate and lay their eggs in their home ponds. Frogs and toad "spawn" can be easily (but carefully) collected and its development watched in an aquarium. Most reptiles lay eggs, leaving them to hatch in sunny, sandy banks or deep in the warmth of a compost heap. In some, the egg hatches inside the mother, so live young are born.

salamander

lizard

newt

frog

toad

Left: *Some snakes are harmless, but never touch a snake without knowing what it is.*

Right: *Frogs have longer, slimmer legs and smoother, damper skins than the dry, warty toad.*

THE LIFE CYCLE OF A FROG

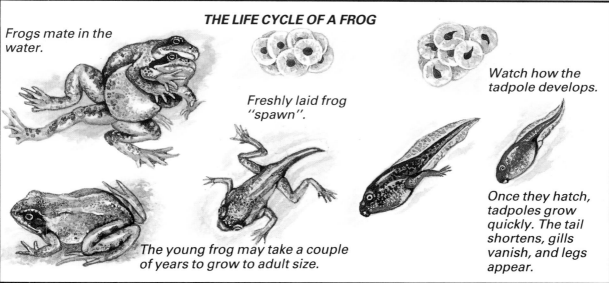

Frogs mate in the water.

Freshly laid frog ''spawn''.

Watch how the tadpole develops.

Once they hatch, tadpoles grow quickly. The tail shortens, gills vanish, and legs appear.

The young frog may take a couple of years to grow to adult size.

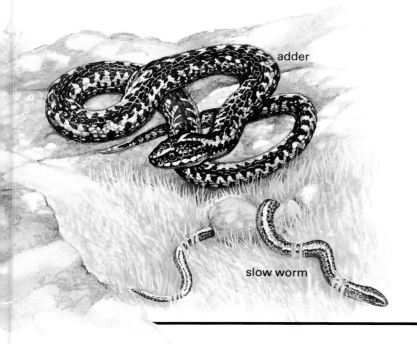

adder

slow worm

AT THE SEASIDE

Rocky coasts, with sandy beaches and cliffs, are ideal for the nature watcher, as there will be something to see almost throughout the year. Always begin at the bottom of the beach and work upwards, to avoid being cut off by the incoming tide. Remember that cliff-climbing is dangerous. It is far better to use your binoculars safely from a distance.

Make a note of where you find seaweeds and shellfish. Were they mainly in the distinct zones between the high- and low-tide marks? Try marking the shells of limpets and whelks, and the spot where you found them, to see if they move about from day to day. Take a close look at the leaves of seashore plants. How different are they from those in the fields inland?

SANDY BEACHES

Any hole or bump on a sandy or muddy beach is likely to have been caused by an animal. Dig quickly – you're sure to find something. Try washing any worms and shellfish out of a spadeful of mud in a sieve (always useful on the beach) and use a lens to identify the various sorts. Look carefully too, among the seaweed and on the sand beneath it. Empty shells should be investigated, as well as the net-like holdfasts of seaweeds, for any animals sheltering there. Sandy bays and muddy estuaries may well be covered in the footprints of ducks, geese, waders and gulls. Use beached boats or dunes as natural "hides" and follow the rules for bird-watching.

marram grows horizontally and vertically, stabilizing dunes

Left: *Sand dunes are difficult places for plants and animals to live in because they are constantly shifting. Marram grass roots hold them together, so that other plants, such as sea buckthorn, can colonize them. You may see terns or oystercatchers nesting near the high-tide mark and rabbits burrowing throughout the dunes.*

SEABIRDS

Seashores are important feeding grounds for many groups of birds gathering in huge numbers on their way to and from their breeding sites. Notice how each species choses a different nest site – puffins in burrows, guillemots on ledges, razorbills under rocks, herring gulls on grassy slopes, kittiwakes on small projecting rocks – often under an overhang.

Watch carefully, especially when there are nestlings to feed and try counting the number of eggs in each clutch to work out an average. Is it higher at the centre or the edge of the colony? How often do the adults feed their young, and does the number of visits differ through the day – or if the weather is bad?

Seabirds are tremendously noisy, but try to pick out the characteristic calls, such as the kittiwake repeating its own name, or the herring gulls hoarse laugh. Look for the young herring gull, pecking at the red spot on its parent's beak – it does not get fed otherwise!

Above: *Guillemots and kittiwakes congregate on cliffs and rocky stacks.*

Below: *Rock pools can offer hours of fun. Always take a guide book, shrimping net, and a keeping jar with you. A hand lens and square-sided transparent box are handy for close observation. Always sit with your face to the sun (to avoid casting a shadow) and wait quietly, avoiding sudden movements.*

Right: *Puffins make a comical sight at breeding time.*

Below: *You may be lucky enough to see a seal in the sea – one of the survivors of the recent epidemic of virus disease.*

DOWN TO THE WOODS

Above: *Today most ancient forest has gone and man has shaped our woodland scenery. But wildlife has adapted well: deciduous woods* (left) *are often richest, but conifers* (above), *especially plantations in the younger stages of growth, can be fascinating.*

Woodland is not the easiest of habitats for the nature watcher. Birds tend to move about up in the canopy, almost out of sight, while mammals nest quietly by day often underground or in a dense thicket. Plants too, because of the shady conditions, are often hard to find.

Try building a rough hide of branches, and waiting patiently. Good spots are near to drinking pools or beside a rotten log – popular because of all the wood chippings around it. Or you can use bait: a few nuts, or some fat pressed into cracks in the tree bark, will tempt many birds. Watch how various members of the same family (the tits for example) share their habitat by feeding in separate zones – blue tits in the canopy, great tits on the ground – and woodpeckers and nuthatches use different tree-climbing techniques.

Above: *Look for holes in trees – they provide homes for many woodland birds.*

IN AND AROUND A TREE

In full leaf, the upper parts of a deciduous tree, an oak for example, is a haven for wildlife. Huge numbers of insects feed upon the leaves, and on each other. The lower leaves can easily be examined: look on or under the leaves and around the leaf stems.

In spring these insects provide food for the adults and young of woodland birds. The upper layers of the trees also provide nesting sites, but it is hard to see the birds while the leaves offer such good protection. Sit quietly behind a shrubby cover, and wait.

Examine tree-bark with a hand lens – you may see a host of tiny spiders, mites and insects crawling about. Lift rotting or peeling bark carefully to reveal boring insects and hidden pupae. Snails, woodlice and millipedes may also hide beneath the bark, and the adults and young of some butterflies and moths may hibernate there. Look, too, for nuts jammed in to crevices by birds and for holes that may be the homes of woodpeckers, nuthatches, tits or owls.

Above: *A noticeable feature of coniferous woodland is the silence. It is essential therefore that you are as quiet as possible. It is worth hiding near a pool of water – cross-bills or other seed-eaters may come to drink from it.*

Left: *Galls are produced by trees as a reaction to a parasitic wasp laying an egg. The growing wasp larva is protected by the gall.*

Below: *Bats should be left undisturbed in their roosting holes.*

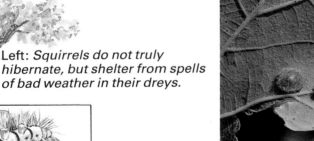

Left: *Squirrels do not truly hibernate, but shelter from spells of bad weather in their dreys.*

Above: *Even a massive oak can sometimes have most of its leaves eaten by winter moth caterpillars.*

Left: *Look for shelf-like bracket fungi on the trunk.*

UP IN THE HILLS

Left: *The "uplands"
start with rough
grazing used only
through the summer.
This grassland merges
into hills and
moorland, often boggy
and with small lakes
and ponds, with
heather and a few
stunted trees. From
the moors rise the real
mountains.*

Below: *Upland
animals have to
cope with severe
conditions. Some
migrate to lower,
richer altitudes
in the winter.*

Hills and mountains are harsh places for the
nature watcher and animals and plants that
live there. Food is often scarce and there may
not therefore be many animals, as many
migrate to lower altitudes.

The uplands are well worth exploring, but
remember that the weather may change very
quickly, so be sure to take all the necessary
safety precautions. Always bear in mind that
what was rugged walking may suddenly
become dangerous. Look always for the
specialists living in these areas, and try to
find the features that enable them to survive.
The flowering plants are often spectacularly
pretty, but note how low-growing they are to
escape the wind, and how short their flowering and fruiting season is. Look for saxifrages
(the name means "rock cracker") and see
how they root in tiny pockets of soil.

Above: *The mountain hare is easily visible before the snow.*

Above: *Scan the horizon for soaring birds of prey. Sunny days are the best – there will be more updraught and the birds will be rising.*

UPLAND BIRDS

In the uplands, small birds are few and far between: seed-eaters like the snow bunting and twite, and insect-eaters like the meadow pipit predominate. In summer, many waders breed on boggy moorland: look for curlew (with a thrilling song-flight), dunlin and golden plover and how well their apparently bright coloured plumage blends with heather and lichens. This is excellent habitat for birds of prey: soaring golden eagles and buzzards, dashing peregrines and merlins, and slow low-flying hen harriers and short-eared owls.

HIGHER AND HIGHER

As you climb higher from moorland up the mountainside, you pass through zones in which the habitats become increasingly severe. There will be less vegetation the higher you go, with stunted trees only occurring in sheltered gullies. Rock cliffs and screes take over and then finally comes the zone where there is almost permanent snow and ice. The snow fields are home to the ptarmigan, white for winter camouflage, brown in summer.

DO'S AND DON'TS OF THE MOUNTAINS

√ take ample food
√ take ample warm waterproof clothing
√ wear one item of bright clothing (to raise the alarm in case of accident)
√ wear proper boots
√ allow ample time for your journey

✕ go without a map
✕ go without telling someone your route
✕ go alone if possible
✕ go without a compass
✕ go without a whistle to raise the alarm
✕ take any risks
✕ stray from marked paths

minnow

brown trout

stickleback

gudgeon

Freshwater habitats are varied and wide-spread. Some support a great deal of life, and are among the most valuable of habitats. Others have virtually no life at all.

PONDS AND LAKES

For many people, watching nature in ponds is the most rewarding activity. Many of the techniques described here can be applied to other freshwater habitats. Wellingtons are a must, but a long-handled net, sieve, dishes and jars, will come in handy. Some of the most exciting things can be seen by sitting patiently at the pond's edge, partly concealed among bushes and reeds. Watch out for waterbirds nesting and for dragonflies perching on waterside vegetation. Frogs may come to bask in their favourite spots and during the summer many of the large species of fish

shoal in the shallows in order to spawn. You may see aquatic bugs and snails – and if your eyes are sharp, water voles and weasels.

Scan the surface of the water for bugs and other insects and look for signs of fish activity. In the water, many species of pond weeds provide both food and shelter for larval and adult insects. Pull some weed out in a net and place it in a tray so that you can inspect it closely.

RIVERS, STREAMS AND CANALS

From source to sea, a river's character changes dramatically and so the plants and animals vary too. Moving water always has something to offer the naturalist, but there will be more species to see in spring and summer. Early morning or evening are the most rewarding times to visit. Look for clues

shrews and watch out for holes in the bank. They may be the homes of water voles, or kingfisher.

Remember never to go near fast or deep water and never lean over a bank. The best places are streams or shallow waters with good visibility where you can approach the water slowly on your hands and knees. Never let your outline be visible against the skyline – fish are particularly sensitive to shadows cast over the water. Polarized sunglasses allow you to see more clearly into the water, as they reduce the glare. Look for any movement: the flash of silver as a fish rolls to catch an insect, or a white diamond shape suddenly appearing as a trout opens its mouth to breathe. But studying fish is difficult without catching them. Use care and an unbarbed hook.

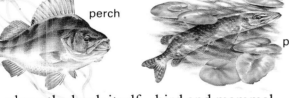

perch

pike

chub

roach

along the bank itself – bird and mammal footprints, droppings, perhaps the signs of a struggle where a mink surprised a duck and carried it off. Listen for the squeak of water

Above and right: Ponds are one of the easiest habitats to observe and understand. Why not try pond dipping? Pass your net along the surface of the water. You'll find pond skaters, whirligig beetles and mosquito larvae. Now swish your net through the weeds for snails. Lastly, check the *bottom for caddisfly larvae and tubifex worms.*

Take home a container of water rich in microscopic life. Your finds will keep in a bucket for a few days, but you could try setting up a proper aquarium. Remember, though, that many creatures are carnivorous and may eat each other.

IN TOWNS AND CITIES

Towns – particularly those with plenty of large gardens and a park and churchyard – are really just extensions of the open country-side. Our garden plants, though often alien in origin, seem quite satisfactory food plants for insects, slugs and snails, and mammals and birds, and provide perfectly satisfactory shelter all year. A number of animals and birds have adapted well even to the "concrete canyons" of city centres, including unexpected or shy species, such as magpies, tawny owls, kestrels, carrion crows and foxes.

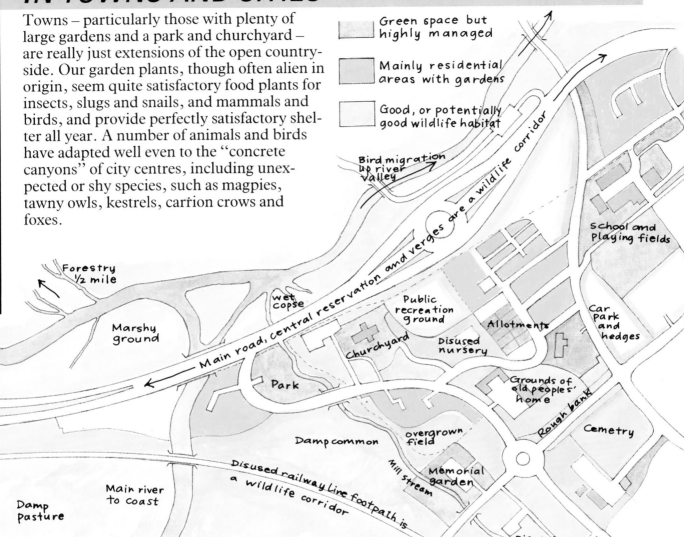

Green space but highly managed

Mainly residential areas with gardens

Good, or potentially good wildlife habitat

Bird migration up river valley

Main road, central reservation and verges are a wildlife corridor

Forestry ½ mile

Marshy ground

Wet copse

Public recreation ground

Churchyard

Disused nursery

Allotments

School and playing fields

Car park and hedges

Park

Grounds of old peoples' home

Rough bank

Cemetry

Damp common

overgrown field

Memorial garden

Mill stream

Disused railway line footpath is a wildlife corridor

Main river to coast

Damp pasture

Wet scrub

Disused yard

Above: *A detailed town map shows the areas most likely to be of interest to nature watchers.*

Natural predators, such as the fox have adapted well to town life, often scavenging from dustbins. They have been joined by a ferocious un-natural predator that does a lot of harm – the cat.

Wild flowers are plentiful on patches of waste ground, particularly thistles, bindweeds and shrubs, such as the buddleia. The purple heads of buddleia attract so many butterflies that it is sometimes known as the butterfly bush.

IN THE PARK

Parks and other open spaces offer plenty of opportunities for wildlife to thrive. Usually more open and less sheltered than most gardens, they usually have fewer flowers, but more trees. Many parks have lakes or ponds, too, and these offer the nature watcher a variety of wildlife, from wintering and breeding birds, flowers and amphibians, to insects, such as dragonflies. Some of the larger parks may support rare, old trees, homes to hole-nesting birds, roosting or breeding bats, wood-boring insects, fungi and other interesting species.

Many insects make good use of the extra warmth, food and shelter provided in our towns and cities. Moths are particularly common and at night turn the buddleia (or butterfly bush) into a moth bush! On the hottest days of summer, black ants make their presence felt, when the queens and winged males emerge, filling the air with

Above left: *Watch to see which plants arrive first on a new habitat. How long is it before they are replaced? Building sites left unworked for a time will soon be covered in seedlings – a feeding area for birds.*

Above: *House martins and swallows originally would have nested on cliffs, or in caves or hollow trees. There are so many of them in our cities today that they have now become prey to kestrels.*

"flying ants". Less welcome still, are wasps, which often nest in holes in banks or walls.

Other insects commonly seen include the pretty and familiar ladybirds and the strikingly large and impressive stag beetle, with its massive "pincers".

Right: *Flies, lacewings, wasps, birds and bats enjoy the shelter of our roofs. Bats should not be disturbed.*

Left: *The nest of the common wasp is surprisingly large and intricate.*

IN THE GARDEN

Your own garden conceals a wealth of wild-life, large and small. Collecting boxes, a lens (and a microscope) will all increase your opportunities to watch closely what is going on. How does a slug slide on its trail of slime? How does a snail eat a leaf – and does its mouth differ from a caterpillar's?

Set up a "worm farm" in a glass-sided box to see how worms pull leaves down into the soil creating valuable humus. Set up an ant colony in a similar box, feeding them with sugar or jam, and watch their busy lives. Outdoors, try to track how far ants travel from their home nest in search of food. What do they bring home?

slug   snail   centipede   woodlouse   caterpillar   earthworm

ENCOURAGING THE BIRDS

Most people first get to know birds through those they see in the park or garden. Setting up a good table will attract masses of birds for you to observe. Be sure to accommodate all tastes on your table. General scraps suit many birds, such as house sparrows and starlings, while a hanging bone may attract a great spotted woodpecker. Peanuts in a hanging basket are ideal for tits and greenfinches, but many other birds have quickly learnt how to feed from them. Fruit, especially half-rotten apples, tempts thrushes and all birds relish water for drinking and bathing. Stop feeding during the spring and summer months.

Above: *Your garden is teeming with wildlife. Bread soaked in milk makes a tasty snack for a visiting hedgehog.*

Below: *Feeding tables and carefully sited nest boxes will encourage many birds – and other wildlife – into the garden.*

Left: *How many different kinds of spider's web can you find? Try looking on a dewy morning.*

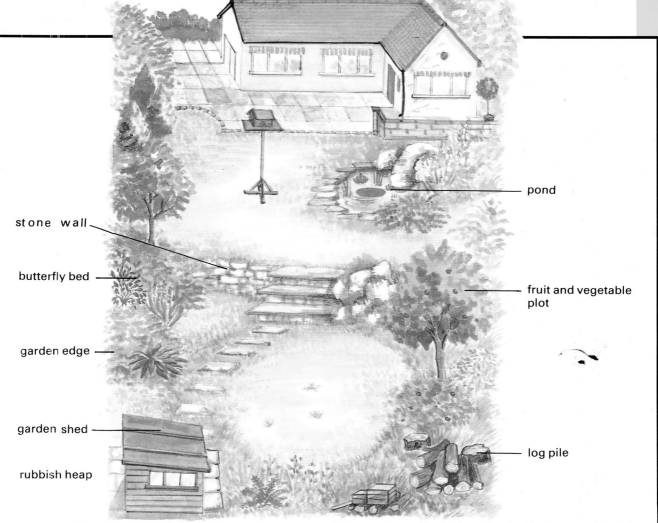

stone wall

butterfly bed

garden edge

garden shed

rubbish heap

pond

fruit and vegetable plot

log pile

CREATING A WILDLIFE GARDEN

Provide a **fruit and vegetable plot** – no wildlife is wanted here if possible, but wood-pigeons and thrushes enjoy it. Allow some nettles to grow along the **garden edge** for butterflies. A **log pile** will attract wood-boring insects, fungi and hedgehogs and a **stone wall** will attract still more insects in the unpointed cracks. A **"butterfly"** bed will require good nectar plants, including *nicotiana* and *sedum*. These will also attract night-flying moths.

A **rubbish heap** will provide an ideal hibernation site for hedgehogs and grass snakes. Remember hedgehogs also burrow into autumn bonfire heaps – so check before anyone lights up. A **garden shed** may be a good nest site for swallows, so leave the window open. And finally, don't forget that garden **pond**, where you could introduce frogs and newts.

Below: *A rough old box, buried in logs and filled with dry grass and leaves, will often tempt a hedgehog to overwinter. Resist the temptation to disturb the sleeping occupant more than a few times.*

KEEPING RECORDS

The notes that you make, and the sketches or photographs that you take on your nature watching expeditions, will be of great help to you as you get to know more about your subject. Just as important, many local or national natural history and conservation societies keep data banks of the occurrence and distribution of animals and plants. This information can be used to compile distribution maps: without the contributions of amateur naturalists like yourself, the distribution maps in field guides could never be produced.

A great deal of information – such as regular counts of birds at birdwatching spots – is used by local or national conservation bodies to defend our countryside from unwanted and ecologically harmful develop-ments. The people involved in this work have to be given regular, updated facts.

THE COUNTRY CODE
√ guard against all risks of fire
√ keep all gates closed
√ keep dogs under proper control
√ keep to the footpaths
√ be careful not to damage hedges, fences or walls
√ take your litter home
√ safeguard water supplies
√ protect all wildlife

Right: *If you come across a developing toadstool, go back and make notes of it as it grows. Your sketches will tell you a lot about the different stages of its life.*

Below: *A Polaroid camera provides you with an instant record of your observation.*

Foxglove (Digitalis purpurea)
Place: Mitcham Common
Date: 2nd June

Left and below: *From sketches in pencil and crayon in your notepad, draw and paint pictures for your records. Remember to note where and when you saw something.*

USING FIELD GUIDES

There are many good field guides on the market that are pocket-sized, reasonably cheap and accurate. Build up a collection of guides on those topics that particularly interest you – your accuracy of identification will improve and your interest in a fascinating hobby will increase even more. Make sure you have discovered how your guide is set out and that you have understood any identification keys it may use. In many guides it is not possible to show, for example, different species of insects – there may only be representatives of the major groups. Sometimes you may need more specialized guides.

Trying to photograph wildlife can be difficult, as it is hard to get close enough for a detailed shot. With a good deal of patience, you should get some fun results!

Left and above: *"Fast" colour films are available, but they give a grainy image for nature photography. It is better to use a slower film and an electronic flash, which helps to give a greater depth of focus and "stopping" movement.*

Right: *Most small plants, particularly parts of flowers, and small animals such as insects, can be studied much better through a good lens – ×10 magnification* *is ideal. Many cameras can take either extension rings or supplementary lenses to create a magnified "close-up" image.*

Index

Published in 1989 by
The Hamlyn Publishing Group Limited
Michelin House, 81 Fulham Road, London SW3 6RB

Copyright © The Hamlyn Publishing Group Limited 1989

ISBN 0 600 56099 6

Printed and bound in Italy
Front jacket illustrations: John Davis, Tony Morris
Front jacket photographs: Eric and David Hosking

Photographs: Heather Angel 2 left and right, 27; Biofotos/Jeremy Thomas 4
left; Bob Gibbons 7, 17 top and bottom, 22, 23 bottom, 25 bottom; Eric and
David Hosking 9/John Hawkins 5 left/G E Hyde 31; Frank Lane Picture
Agency/Ray Bird 27 bottom/Roger Tidman 10/Roger Wilmshurst 18; Natural
Image/John Howard 28/Julie Meech 19 top/Peter Wilson 27 top right; Nature
Photographers 19 bottom left/N A Callow 13 centre/Hugh Clarke 26/Andrew
Cleave 19 bottom right, 20 left, 24–25/E A Janes 6/Hugh Miles 23 top/Owen
Newman 4 right/Paul Sterry 13 top, 15 left and centre; The Octopus Group
Picture Library 5 right, 13 bottom, 15 right, 20 right, 21; Judy Todd 12.
Illustrations: Priscilla Barr, Tony Morris – Linda Rogers Associates,
Richard Geiger, Mei Lim, Gill Tomblin
Design: Mei Lim